THE CHAOZHOU BOOK OF TEA
Translation and editorial matter © 2026 by Simone Chiatante
Original Title: *Chaozhou Cha Jing* by Weng Huidong
All rights reserved

The source text is an unabridged republication of the work
originally distributed in handwritten copies. Published by
courtesy of the Shantou Library.

Translation: Simone Chiatante 勇和
Cover Calligraphy: Xiaoge Wang 王晓戈
Graphic Design and Illustrations: Martin Pezer

Chinese Chapbooks: Volume 2
Published by Subway Dharma Collective

Weng Huidong

翁辉东

The Chaozhou Book of Tea
潮州茶经

Translation by Simone Chiatante

Subway Dharma Collective

CONTENTS

p. 7 **Introduction**

13 **The Chaozhou Book of Tea**

19 **The Meaning of Gong Fu Tea**

23 **The Essential Nature of Tea**

27 **Water Sourcing**

31 **A Lively Fire**

35 **Teaware**
The Teapot
The Lidded Bowl
The Teacup
The Tea Basin
The Tea Tray
The Tea Mat
The Water Jug
The Water Basin
The Dragon Jar
The Red Clay Stove
The Sand Kettle
The Feather Fan

71 **Preparation**
    Preparing the Utensils
    Sorting the Tea
    Observing the Water
    Pouring Water
    Scraping Off the Foam
    Drenching the Pot
    Heating Up the Cups
    Steeping the Tea

87 **Translator's Note**

Introduction

The role of tea preparation at the center of a people's cultural life has been the subject of multiple essays written by East Asian scholars, especially at times when East and West had their most intense interactions. The renowned "Book of Tea" by Okakura Kakuzo (1863–1913) had been written directly in English, targeting a Western audience on the main aesthetic and philosophical aspects of Japanese and Asian cultures, and taking tea practice as an exemplary art that, indeed, condensed them all.

Half a century after the publication of Kakuzo's essay, Weng Huidong, a passionate scholar and educator from Chaozhou, in the south of China, attempted a similar endeavor. Chaozhou is part of what nowadays is called the area of Chaoshan, a name that not only designates a geographic region, but also a proudly preserved cultural and linguistic identity. Here tea practice does not only stand for the aesthetic cultivation of an elite of intellectuals. As Huidong emphasizes in an unpublished preface to the essay here presented, all the people of Chaozhou infuse every moment of their daily life with elegance, everywhere from home to the town's walkways, even to their workplace, through their tea drinking. Setting up their specific utensils at a table, they make their customary pot and three little cups the center of their social life. A solitary meditation at times, or a habitual, almost ceremonial gathering akin to what coffee time represents in some Mediterranean countries. It is the Gong Fu tea practice that has become popular all over the world and that is attracting people in the West who prioritize a more mindful and organic approach to consumption, rather than a consumeristic one. It is also what became the preferred style of tea brewing all over China in the last decades, anywhere the aesthetic and meditational aspects of this ancient art still hold greater significance.

Written in 1957, The "Chaozhou Book of Tea" showcases a cultivated language, terms and metaphors that take on the millenarian Chinese arts and history. The style is elegant, yet Wang's is a modern perspective. While his most famous predecessor in Chinese tea practice dissemination, Lu Yu (733–804) with his "Classic of Tea" (茶经), belonged to an era where tea was processed, sourced, and prepared in completely different manners, Weng Huidong meticulously describes the actual ways and utensils employed in the 1950s in Chaoshan, and that are still in use. While nowadays much attention is paid by tea aficionados to the exact water temperature degrees, grams of tea leaves per mL of water, precise calculation of steeping times for tea-type and so on, he focuses on each step organically, indeed with the same attitude of the living tea masters in Chaoshan, who instead stress on the importance of preparing tea as naturally as possible according to the circumstances. Steeping time should be moderated by preventing bitterness and astringency, the pot should be small, and the quantity of leaves in the pot should be determined by eye. As simple as that. Assessing the quality of a pot, on the other hand, is assimilated to the understanding of ancient Chinese clepsydras or to knowing the value of jade artifacts. And it is in these passages that we see that tea is not a simple commodity or a mere local product, but part of a long local tradition, perhaps at Weng's time still obscure to the rest of the country.

Our aim, thus, is to take the author's intention of spreading and clarifying the essence of Chaozhou culture and Gong Fu tea etiquette to a wider audience, and make this first essay on the subject international, by providing all English-speaking readers and tea researchers Weng Huidong's cultivated insight.

THE CHAOZHOU BOOK OF TEA

潮州茶经

人类嗜茶，殆与酒同。以为饮料，几遍世界。原因茶含单宁酸，具激刺性，能令人启迪思虑。更有文人高士，借为风雅逸致，凡在应酬交际，一经见面即行献茶。在商业方面，亦赖茶为重要之输出品，揆之事实，茶与人类生活非但占重要性、以为饮料，已属特别；惟我潮人，独擅烹制，用茶良瘾，争奢夺豪，酿成"工夫茶"三字，驰骋于域中，尤为特别中之特别。良辰清夜，危坐湛思，不无念及此杯中物，实具有特别之素质与气味在。

Humanity's fondness for tea is nearly the same as for wine; tea is indeed a beverage that has been spread all over the world. This is because it contains tannins, which work as stimulants. Drinking it can inspire thought, and scholars, literati consume it as an expression of elegance and refinement by offering tea at every social gathering. Evidence shows that business also relies on tea as a key export. Furthermore, tea is not only crucial to various aspects of human life but is also a peculiar beverage.

Only the people of Chaozhou, with their skills in preparing and in sourcing tea of varying quality, have uniquely excelled in perfecting this art, and have distilled it into the three words: Gong Fu Tea[1], which

---

[1] Gong Fu Cha (工夫茶) denotes the meticulous dedication into tea making and preparation. It is often used interchangeably with 功夫茶, pronounced the same way, where *gong fu* (功夫) are the same characters that stand for Chinese martial arts, implying skills

have become renowned across the country as the art has achieved exceptional character among the exceptional. On fine nights, while sitting in deep contemplation, I also cannot help but concentrate on the intrinsic essence and aroma of the tea in my cup.

---

and accomplishment. The 7th edition of the Commercial Press Modern Chinese Dictionary combines 工夫茶 and 功夫茶 into a single entry. However, "The Dream Factory Collection: The Romance of Chaozhou and Jiaxing" by Yu Jiao (1751–?) contains the earliest record of Gong Fu tea art, expressed by the characters 工夫茶.

工夫茶之特别处,不在于茶之本质,而在于茶具器皿之配备精良,以及闲情逸致之烹制。

潮地邻热带,气候常温,长年需饮以备蒸发。往昔民安物泰,土地肥美,世家巨族野老诗人,好耽安逸,群以饮茶相夸尚,变本加厉。对于"茶质"、"水"、"火"、"用具"、"烹法",着着研求,用于陶情悦性,消遣岁月。继则不惜重资,购买杯碟,已含弄骨董性质。所以工夫茶之驰誉域中,其原因多也。钱塘陈坤子厚,咏工夫茶诗云:"何人曾识赵州来,品到茶经有别裁。不咏卢仝诗七碗,金茎沆露只闻杯"。

The meaning of Gong Fu tea lies not in the essence of the tea itself, but in the masterfully crafted utensils, and in the serene, mind-liberating preparation. Chaozhou is situated near the tropics and its climate is warm, demanding that its people drink all year-round to prevent dehydration. In the past, when the people enjoyed peace and prosperity, and the land was fertile, aristocratic families or even feral poets could indulge in leisure and enjoy appreciating tea together. They became more and more serious, meticulously studying the quality of tea, the water, the fire, various utensils, and the brewing methods, in order to soothe their spirits and pass the time. They spared no expense in purchasing cups and saucers, in a manner that resembles the collecting of antiques. That is how the fame of Gong Fu tea has spread all over the country and even beyond. In a poem about Gong Fu tea, the poet Chen Kun Zihou[2] from Qiantang wrote: "Only those who have known Zhaozhou[3] have savored the unique taste of the 'Classic of Tea[4]'. There is no need to recite Lu Tong's 'Seven Bowls of Tea[5]'; dew gathers on its golden stem, whirls aroma from the cup."

---

[2] Chen Kun (1821–?), also known by his courtesy name Zihou (子厚).

[3] Zhaozhou (赵州), an eminent Buddhist monk of the Tang dynasty, known for his famous koan about a "cup of tea".

[4] 茶经, by Lu Yu (733–804), recognized as the first monograph on tea ever written. Sometimes, it is also translated as the tea "Canon" or "Sutra".

[5] 七碗诗, by Lu Tong (790–835), perhaps the most famous poem about tea, where the poet clearly attributes to tea drinking health properties, as well as its spiritual and intellectual bearings.

爱将工夫茶之构造条件胪列如下:

茶之本质。

我国产茶名区,有祁门、六安、宁州、双井、弋阳、龙井、太湖、武夷、安溪,以及我潮之凤凰山、待诏山等。而茶之制法,则有红茶、砖茶、绿茶、焙茶、青茶等。茶之品种,则有碧螺春、白毛猴、铁观音、莲子心、老鸟嘴、奇种乌龙、龙井等。潮人所嗜,在产区则为武夷、安溪,在泡制法则为绿茶、焙茶,在品种则为奇种、铁观音。

Below, I will clearly outline the foundations of Gong Fu tea:

The Essential Nature of Tea.

Famous tea-producing regions in our country include Qimen, Lu An, Ningzhou, Shuangjing, Yiyang, Longjing, Taihu, Wuyi, and Anxi, as well as Fenghuang Shan and Daizhao Shan[6] in our area of Chaozhou. Regarding tea processing methods, we have: Red Tea, Brick Tea, Green Tea, Roasted Tea and Fresh Tea[7], among others. Tea varieties include Bi Luo Chun, Bai Mao Hou, Tie Guan Yin, Lian Zi Xin, Lao

---

[6] A mountain connected with Fenghuang Shan, where grows a variety of tea known as Xianshui (仙水). This tea was listed as a tribute to the imperial court during the Hongzhi reign of the Ming Dynasty.

[7] qing cha (青茶), another traditional way to name today's Oolong tea, specifically those that have been processed lightly. The reference here is to heavily and lightly roasted types of Oolongs.

Niao Zui, Qi Zhong Wu Long[8], and Long Jing. The people of Chaozhou are particularly fond of tea produced in Wuyi and Anxi areas, of Green Tea and Roasted tea in terms of processing methods, and of Qi Zhong and Tie Guan Yin in terms of varieties.

---

[8] In this case, *wulong* (乌龙) has been kept in Pinyin rather than rendered with its internationally renowned transcription *Oolong*, because it refers to a specific tea cultivar, and not to the processing method Oolong is known for today.

取水。

评泉品水,陆羽早著于先。潮人取水,已有所本,考之《茶经》:"山水为上,江水为中,井水其下。"又云:"山顶泉轻清,山下泉重浊,石中泉清甘,沙中泉清冽,土中泉浑厚;流动者良,负阴者胜。山削泉寡,山秀泉神,置水无味。"甚且有天泉、天水、秋雨、梅雨、雪水、敲冰之别,潮人嗜饮之家,得品泉之神髓,每有不惮数十里,诣某山坑取水,不避劳云。

Water Sourcing.

Lu Yu was the earliest scholar to discuss water sourcing and appreciation. Therefore, the people of Chaozhou already had a basis for water sourcing. The renowned "Classic of Tea[9]" states: "Mountain water is superior, river water is mediocre, and well water is the lowest." It also states: "The springs from mountaintops are light and clear, the springs from the mountain bases are heavy and turbid, springs from amidst rocks are clear and sweet, springs from sand are clear and rough, and springs from soil are thick and muddy. The flowing springs are good, those from shaded areas are superior. Springs from steep mountains are rare, those from beautiful mountains are divine, yet stagnant water is tasteless." There are even distinctions between natural water sources, water from the sky, autumn rain, rain

---

[9] See note 4.

collected from the plums[10], snow water, and water collected by breaking ice. Since the people of Chaozhou love drinking, they have mastered tasting the essential characters of springs, and do not mind traveling long distances to collect water from specific mountains and pits, regardless of any effort.

---

[10] Perhaps referring to water collected from plum flowers.

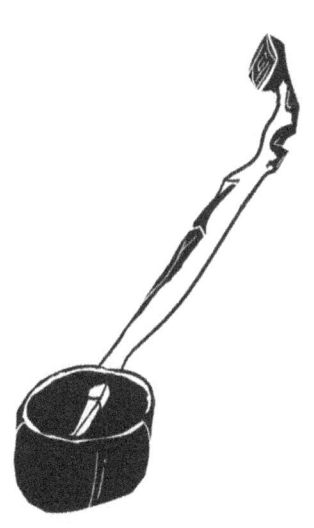

活火。

煮茶要件,水当先求,火亦不后。苏东坡诗云:"活水仍须活火烹"、活火者谓炭,炭之有焰也。潮人煮茶多用绞只炭,以坚硬之木,入窑室烧木脂燃尽,烟嗅无存,敲之有声,碎碎莹黑,以之熟茶斯为上乘。更有橄榄核炭者,以乌榄剥肉去仁之核,入窑室烧,逐尽烟气,俨若煤屑,以之烧茶,焰活火匀,更为特别。他若松炭、杂炭、柴、草、煤等,不足以入工夫茶之炉矣。

A Lively Fire.

Among the essential elements of tea brewing, water is prime, but fire does not come last. Su Dongpo wrote in a poem, "Before lively water comes a lively fire.[11]" "Lively fire" actually refers to a flame nourished in charcoal. The people of Chaozhou often prepare their own charcoal, which is made from hardwood heated in a kiln. The wood resin burns completely away, leaving no smell of smoke. When struck, it makes a specific sound, and when broken, it shows a glossy black luster. This kind of coal is considered the best quality for tea preparation. Charcoal made from olive pits is also considered very useful. Olive pit charcoal is made from black olives; the flesh is removed, and the pit is extracted to be burned in a kiln, exhausting all

---

[11] The author refers to the first line of the poem 汲江煎茶 "Simmering Tea with Fresh River Water", by Su Shi (1037–1101), also known as Su Dongpo.

the smoke, and leaving only coal dust. Using this charcoal for tea brewing creates a lively, even flame, with its own special character. Other types of charcoal, such as pine charcoal, miscellaneous charcoal, straw, or combining wood and charcoal, are not suitable for Gong Fu tea brewing.

茶具。

《云溪友谈》云:"陆羽所造条器,凡二十四事。"茶具讲究,自古已然。然此只系个人行为,高人逸士,每据为诗料,难言其普遍。潮人所用茶具,大体相同,不过以家资有无,精粗有别而已。今将各家饮茶所常备之器皿列下:

Teaware

In "Amical Discussions in the Clouded Valley[12]", we read: "Lu Yu provides a list of twenty-four utensils altogether." Teaware has been carefully selected since ancient times, but choice is always bound to personal preferences. This has often been the subject of poetry by literati or scholars, and it has been difficult to standardize. The teaware used by the people of Chaozhou though, is generally always the same. Differences consist only of how refined or coarse each piece is, depending on the wealth of the household who owns it. Here is a list of the utensils that can commonly be found in various households:

---

[12] 云溪友谈, by Fan Fu (?–?), a collection of records from the middle to the late Tang dynasty.

茶壶。

俗名冲罐，以江苏宜兴硃砂泥制者为佳。其制肇于金砂寺老僧。而潮人最珍贵者，为孟臣、铁画轩、秋圃、萼圃、小山、袁熙生等。壶之样式甚多新颖。即如壶腹款式，运刀刻字，亦在乐毅黄庭之间，人多宝贵之。

壶之采用，宜小不宜大，宜浅不宜深，其大小之分，更以饮茶人数定着。爰有二人罐、三人罐、四人罐之别。其深浅则关系气味，浅能酿味，能留香，不蓄水，若去盖浮水，不颇不侧，谓之水平。覆壶而口嘴提柄皆平谓之三山齐。壶之色泽有朱砂、古铁、栗色、紫泥、石黄、天青等，间有银朱闪烁者，乃以钢朱和制之，珠粒累累，俗谓之柚皮砂，更为珍贵，价同拱璧，所谓硃土与黄金争价，即指此也。壶之款式，有小如蜜柑者，有瓜形、柿形、菱形、鼓形、梅花形，又有六角形、栗子、圆珠、莲子、冠桥等。式样精美，巧妙玲珑，饶有风趣。

The Teapot.

Teapots are commonly known as "infusion pots[13]". The best of them is made of red clay from Yixing, in Jiangsu. Their manufacturing began with the old monks of Jinsha Temple. The pots that are most valued by the people of Chaozhou are those created by such masters as Meng Chen, Tie Hua Xuan, Qiu Pu, E Pu, Xiao Shan, and Yuan Xi Sheng[14], among others; but many new styles of pots have also been created. Even the inscriptions carved on some of the teapots' bellies are inspired by the works of Yue Yi[15] and Huang Ting[16], increasing their value. Teapots should be small rather than large, and shallow rather than deep. Their

---

[13] *chong guan* (冲罐), perhaps to distinguish them from kettles.

[14] Names of masters and studios which crafted teapots in the area of Chaoshan.

[15] 乐毅, perhaps referring to the Lord of Changguo, a military leader during the Warring States period (c. 475–221 BC).

[16] Huang Tingjian (1045–1105), a renowned calligrapher, painter, and poet from the Song dynasty.

size should be determined by the number of people drinking tea, the most popular being those for two, three or four people. The depth of a teapot directly influences taste; a shallow surface allows brewing the tea's essential flavor and aroma without retaining too much water. If, upon removal of the lid, the pot floats on water without tilting nor leaning to one side, we can call this pot "Shui Ping[17]". If the pot's opening, the spout, and handle are all aligned, we call it a "three-mountains level" teapot. There are several combinations of the teapot's color and texture, among which can be named the cinnabar type, the antique

---

[17] These assessing criteria for evaluating a teapot originally derive from the calibration of ancient Chinese clepsydras, used as water clocks. By ensuring that the water is calm or balanced (水平), it is possible to assess the stability and balance of the pot. Therefore, uniform water flow can guarantee a controlled timekeeping during steeping and pouring. A specific kind of pot design, known for lightness and optimal performance is considered the best and it has been named Shui Ping (水平).

iron type, the chestnut type, the purple clay type, the orpiment type, or the azure sky type. Some types showcase on the surface a shimmer of silver and vermilion. This is due to a mixture of grains of iron and cinnabar in the clay, commonly known as "pomelo peel clay." These are even more precious and can be as valuable as an extraordinary jade treasure[18]. The saying that cinnabar soil competes with gold refers exactly to this. Pot's design varies in size and shape, from as small as a tangerine to as large as an orange; there are pots shaped like melons, persimmons, diamonds, and drums. There are also some modeled on plum blossoms, some hexagonal, some like chestnut, some all-round, some like lotus seed, or decorated like

---

[18] *gong bi* (拱璧), a precious jade disk, in ancient times symbolizing wealth and power. Here, the expression is hyperbolic and emphasizes the rarity of this kind of clay.

crested bridges. The designs are refined, ingenious, and full of charm.

盖瓯。

形如仰钟，而有上盖，下置于垫，俗名茶船，本为宦家贾客自斟之器，潮人也采用之。或者客多稍忙，故以之代冲罐，为其出水快也。惟纳茶之法必与纳罐相同，不能颠预。其逊于冲罐者，因瓯口阔不能留其香，或因冲罐数冲之后，稍嫌味淡，即将余茶掏于瓯中再冲，备饷多客权宜为之，不视为常规也。

The Lidded Bowl[19].

Lidded bowls are vessels shaped like an upturned bell, with a lid and a saucer underneath, commonly known as a "tea boat". Originally used in the imperial officers' households or by merchants on their own, lidded bowls have also been adopted by the people of Chaozhou. Sometimes, when there are many guests and the circumstances are busy, these bowls are used instead of a teapot as they can rinse and pour the tea faster. However, the way of serving tea with them ought to be the same as using a teapot. No carelessness is allowed. This vessel is inferior to a teapot because the bowl's wide mouth does not retain much of the tea's fragrance. Sometimes, if after several steepings, the tea flavor has weakened, more leaves can be added

---

[19] A *gai ou* (盖瓯) is in Chaoshan what is more commonly referred to as *gai wan* (盖碗).

into the bowl and steeped again to serve more guests[20]. However, this can be done out of convenience, and it is not to be considered standard practice.

---

[20] By adding up fresh leaves over the already brewed ones results easier and faster in a Gai Wan; since by being a bowl, its opening is relatively wider than a standard teapot used for Gong Fu brewing.

茶杯。

茶杯若深制者为佳，白地蓝花，底平口阔，杯背书"若深珍藏"四字。此外仍有精美小杯，径不及寸，建窑白瓷制者，质薄如纸，色洁如玉，盖不薄则不能起香，不洁则不能衬色。此外四季用杯，式样有别，春宜牛目杯，夏宜栗子杯，秋宜荷叶杯，冬宜仰钟杯。杯亦宜小宜浅，小则一啜而尽，浅则水不留底。（近人取景德制之喇叭杯，口阔脚尖，而深斟必仰首，数斟始罄，又有提柄之牛乳杯，均为讲工夫茶者所摒弃）。

The Teacup.

The most refined teacups are those of Kushen manufacture, decorated with blue floral patterns on a white background, with a flat bottom and a wide brim, inscribed with the words "Ruoshen Treasure[21]". There are also exquisite small cups, less than an inch in diameter, produced by other kilns in *blanc de chine* white porcelain, as thin as paper and as clean as jade. Indeed, a thick surface could not retain much fragrance, and a clean coloration does enhance the color of the liquor. Furthermore, each season may be

---

[21] These seem to be the cups that set the model for any later cup used in Gong Fu tea preparation. Perhaps they are marked after the name of a Ruoshen (若深) kiln, although different theories are in circulation. The name may have been of a gentleman who first commissioned this kind of cups, or of the artist who first created them. In any case, the mark is believed to have appeared during the Kangxi reign (1661–1722), in the Qing Dynasty. These cups indeed reflect the features described afterwards in the text, but of finest manufacture. Later imitations can be identified by a somewhat watery effect in the underglaze, or other characteristics.

matched with a different type of cup[22]. Ox-eye type cups are suitable for spring, chestnut cups for summer, lotus leaf cups for autumn, and inverted bell cups for winter. A proper cup is also small and shallow; small enough so that the tea could be drunk in one sip, shallow enough so that no water is left at the bottom. (Today, people use trumpet-shaped cups manufactured in Jingde[23], with a wide mouth and pointed bottom, and need to tilt their head back to drink properly, so that it takes several sips before the cup is fully emptied. There are also cups with handles

---

[22] In old Chaoshan tea etiquette, the shape, size, and material of a cup are important for enhancing or concentrating aroma, dissipating, or holding heat. Therefore, some types of cups result suitable for the temperature or for the features of the different seasons.

[23] Jingdezhen (景德镇), in Jiangxi Province, where curiously, also the first porcelain cup type mentioned in this section is famously produced.

like those used for drinking milk, all of which are shunned by those who practice Gong Fu tea.)

茶洗。

茶洗形如大碗,深浅式样甚多,贵重窑产,价也昂贵。

烹茶之家,必备三个:一正二副。正洗用以浸茶杯,副洗一以浸冲罐,一以储茶渣暨杯盘弃水。

The Tea Basin.

Tea basins are shaped like large bowls, and whether deep or shallow, can come in a variety of different styles. Renowned kilns produce very valuable ones. Every tea-drinking household shall have at least three: a main one and two secondary basins; the main basin used to soak teacups, while the other two can be used to soak the teapot and dispose of the tea leaves residues and water from all the vessels used.

茶盘。

茶盘宜宽宜平,宽则足容四杯,有圆如满月者,有方如棋枰者。底欲其平,缘欲其浅。饶州官窑所产素瓷青花者为最佳,龙泉白定次之。

The Tea Tray.

A tea tray should be wide and flat, wide enough to hold four cups. Some trays are round like a full moon, while others are square like a chessboard. The bottom should be flat, and the rim shallow. Trays produced in plain Qinghua blue and white porcelain from the Raozhou [24] official kiln are considered the best, followed by those produced in white porcelain from Longquan[25].

---

[24] A historic region, where the city of Jingdezhen is located, renowned for its Song dynasty porcelain ware, especially the Qinbai (青白) ware, characterized by a distinctive blue-green glaze. Ware produced in the Raozhou region was commissioned for the Chinese court and for overseas markets and delivered to the city of Jingdezhen.

[25] A coastal city in Zhejiang province, famous for its ware produced with a distinctive glaze, in particular its celadon pottery.

茶垫。

如盘而小,径约三寸,用以置冲罐、承沸汤。式样夏日宜浅,冬日宜深,深则可容多汤,俾勿易冷,茶垫之底,托以垫毡毯,以秋瓜络为之,不生他味;毡毯旧布,剪圆形,稍有不合矣。

The Tea Mat.

Tea mats are like trays, but smaller; about three inches[26] in diameter. They can be used to place tea pots while steeping. Light ones shall be used in summer, and thick ones in winter[27]. Thick ones can retain more tea spilled in excess, so that the surface at the bottom does not need to be washed, and the tea mat can be supported by another felt mat. Dried custard melon fiber[28] can also be used as a mat, as it does not retain any odor, but felt mats cut round from an old rag are not suitable.

---

[26] A traditional Chinese *cun* (寸) is slightly longer than a standard English inch, measuring approximately 1.3 inches (3.3 centimeters).

[27] The thickness of the mat determines the heat dissipation from pots and cups, making them more or less suitable for summer or winter.

[28] *qiu gua* (秋瓜), a type of melon whose fiber, once dried, can be prepared for several uses. For instance, it can be pressed or woven to make mats and coasters.

水瓶。

水瓶贮水以备烹茶,瓶修颈垂肩,平底,有提柄,素瓷青花者佳。有一种形似萝卜樽,束颈有嘴饰以螭龙,名螭龙樽(俗称钱龙樽)。

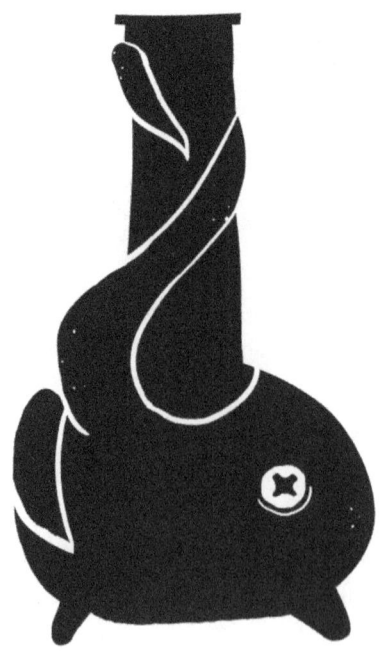

The Water Jug.

Jugs can be used to store water for brewing tea. They usually have a narrow neck, drooping shoulders, a flat bottom, and a handle. Jugs produced in plain Qinghua blue and white porcelain are considered best; A type of jug called a "Chilong Goblet[29]" has its narrow neck decorated with a dragon (also referred to as the "Qianlong Pagoda").

---

[29] The Chilong Goblet (螭龙樽) is an ancient Chinese ritual wine vessel adorned with patterns of *chi* (螭), a mythical hornless dragon from Chinese legends. Often depicted as a young, nascent dragon or a creature of aquatic nature, the *chi* symbolizes power, good fortune, and divine protection. The term *zun* (樽) refers to a type of large, broad-bodied bronze ritual wine jar from ancient China, characterized by a wide mouth and a circular base.

水钵。

水钵为瓷制款式。修颈置于茶床之上,用于贮水,掬以椰瓢。有红金彩者,明代制物也,用五彩金釉,描金鱼二尾于钵底,水动时则金鱼游跃,稀世奇珍也。

The Water Basin.

Mostly made of porcelain and coming in a variety of styles, water basins are placed on the tea table to hold water, which is then scooped with a coconut ladle[30]. Some, made of red gold, date back to the Ming Dynasty. They are decorated with two goldfish at the bottom, glazed in five metals, which seem to swim and leap as the water is stirred, making these basins a rare treasure in this world.

---

[30] A ladle crafted from a coconut shell.

龙缸。

龙缸可容多量坑河水，托以木几，置之斋侧，素烧青花，气色盎然。有宣德年制者，然不可多得。康、乾年间所产，亦足见重。

The Dragon Jar.

Dragon jars can hold large amounts of water from wells or rivers and can be supported by wooden tables. They are usually placed on one side of the tearoom. Those made of plain Qinghua porcelain showcase patterns full of vitality. Some were made during the Xuande[31] reign, but they are rare and difficult to come by. However, those produced during the Kangxi and Qianlong[32] periods are also highly valued.

---

[31] The Xuande (宣德) emperor (1399–1435) of the Ming dynasty.
[32] The Kangxi (康熙) emperor (1654–1722), and the Qianlong (乾隆) emperor (1711–1799) of the Qing dynasty.

红泥火炉。

红泥小火炉,古用以温酒,潮人则用以煮茶,高六七寸。有一种高脚炉,高二尺余,下半部有格,可盛榄核炭,通风束火,作业甚便。

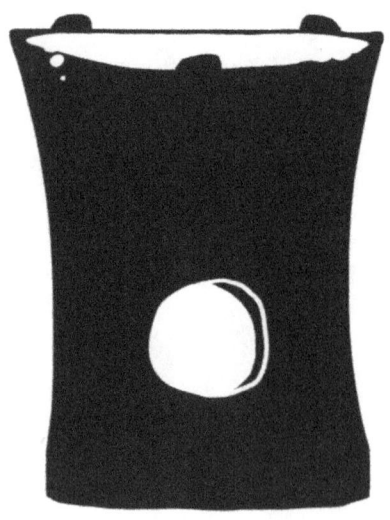

The Red Clay Stove.

Small stoves, made of red clay, were used in ancient times to warm wine, and now they are used by the people of Chaozhou to prepare tea. They are six or seven inches tall, but one kind of taller stove can even be over two feet, with a compartment in the lower half to hold olive-pit charcoal. They are ventilated for fire-control, making them very convenient to use.

砂铫。

砂铫俗名茶锅仔。沙泉清冽,故铫必砂制。枫溪名手所作,轻巧可喜。或用钢铫、锡铫、铝铫者,终不免生金属气味,不可用。

The Sand Kettle.

Sand kettles are commonly known as "Cha Guozi". Sand spring water is clear and fresh in essence; therefore, water kettles ought to be manufactured from local sand[33]. Those made by the famous master Fengxi are light and pleasing to use. Pots made of steel, tin, or aluminum will inevitably leave a metallic taste and cannot be used.

---

[33] This passage implies that, by boiling local water, a kettle made of local sand clay would match the water's physical characteristic best.

羽扇。

羽扇用以扇炉。潮安金砂陈氏有自制羽扇,拣净白鹅翎为之,其大如掌,竹柄丝坠,柄长二尺,形态精雅。炉旁必附铜箸一对,以为钳炭、挑火之用,烹茗家所不可少。

此外茶罐锡盒,个数视所藏茶叶种类而定,有多至数十个者,大小兼备。名贵之茶须罐口紧闭。潮阳颜家所制锡器,有闻于时。又有茶巾,用于净涤器皿。竹箸,用于箝挑茶渣。茶桌,用以摆设茶具。茶担,可以装贮茶器。春秋佳日,登山游水,临流漱石,林壑清幽,呼奚童,肩茶担,席地烹茗,啜饮云腴,有如羲皇仙境。"工夫茶"具,已尽于此,饮茶之家,必须一一毕具,方可称为"工夫";否则牛饮止渴,工夫茶云乎哉。

The Feather Fan.

Feather fans are used to fan the flame in the stove. Jinsha Chen, in Chaoan, crafts his own feather fans from clean white goose feathers, large as a palm, with bamboo handles and silk pendants. The fans are two feet long altogether, exquisitely and elegantly patterned.

By the side of the stove, it is also necessary to keep a pair of chopsticks made of copper, used as tongs for arranging charcoal and stirring up the fire, which is indispensable in brewing tea. There are several kinds of tin canisters, depending on the variety of teas stored; there can be dozens, both large and small. Fine tea needs to be stored in tightly sealed canisters. The tinware crafted by the Yan family in Chaoyang holds a time-honoring reputation. Tea towels can be used to clean all the utensils. Bamboo chopsticks are used to scoop up tea leaves. Tea tables can also be used to display tea utensils. Tea baskets are used to carry the tea utensils.

On fine days in spring or autumn, you can go hike up mountains, go for a swim, refresh in the flowing waters by the rocks. The woods and the ravines are pure and mysterious. A child could be called to carry a tea basket. Setting up on the ground to brew tea, sipping on the rich clouds, it feels like being in an enchanted kingdom. This is what it means to use the Gong Fu tea utensils to the fullest. A tea-drinking household should have all the necessary utensils to be able to call their brewing Gong Fu. Otherwise, it is just thirst-quenching, how could it be called Gong Fu tea?

烹法。

茶质、水、人、茶具,既一一讲求,苟烹制拙劣,亦何能语以工夫之道?是以工夫茶之收功,全在烹法。所以世胄之家,高雅之士,偶一烹茶应客,不论洗涤之微,纳洒之细,全由主人亲自主持,未敢轻易假人;一易生手,动见偾事。

治器。

泥炉起火,砂铫掏水,煽炉,洁器候火,淋杯。

Preparation.

The quality of tea, the water, the participants, and the teaware all require meticulous consideration. If the preparation is careless, how could it be called Gong Fu? The ultimate accomplishment of Gong Fu tea lies entirely in the way preparation is carried. Therefore, when noble households or refined scholars prepare tea for entertaining guests, the host personally oversees every aspect, from washing to serving, never delegating responsibilities to others. Anything handled superficially could lead to failure.

Preparing the Utensils.

Prepare the clay stove, scoop water into the sand pot, clean the stove, adjust the heat, and rinse the cups.

纳茶。

静候砂铫中有松涛飕飕声,泥炉初沸哭起鱼眼时(以意度之,不可掀盖看也),即把砂铫提起,淋罐淋杯令热,再将砂挑置炉上。俟其火硕(老也,俗谓之硕)一面打开锡罐,倾茶于素纸上,分别粗细,取其最粗者,填于罐底满口处,次用细末,填塞中层,另以稍粗之叶,撒于上面,谓之纳茶。纳不可太饱满,缘贵重茶叶,嫩芽紧卷,舒展力强,苟纳之过量,难容汤水,且液汁浓厚,味带苦涩,约七八成足矣,神明变化,此为初步。

Sorting the Tea.

Wait quietly for the sound of wind blowing through pine trees, coming from inside the sand kettle. The clay stove is beginning to boil the water, which starts to swell into little *fish-eyes* (determine this by your senses; do not remove the lid to see), then lift the sand kettle. Warm the pot and the cups by pouring water in them and place the kettle on the stove again. Wait until the boil has "matured"[34] (that is a customary expression to indicate it has grown strong enough), open the tin can and pour the tea leaves onto a piece of plain paper, separating the large from the fine ones. Sort and separate large leaves from fine ones and place them at the bottom of the teapot; place the finest powder in the middle and add the largest tea leaves on top. By placing the tea leaves on top, we can say we are

---

[34] *huo shuo* (火硕), in local Chaoshan dialect indicates that the water has been properly boiled and it is ready for the brew.

"layering the tea". It is better not to fill the pot too much, as valuable tea is made of tender buds that have been tightly rolled and expand vigorously when they open up. If you fill it too much, it won't be easy to drink, the liquor will be thick and bitter. About 70% to 80% of the pot's capacity is sufficient. This is the beginning of a divine transformation.

候汤。

《茶谱》云:"不藉汤熏,何昭茶德。"《茶说》云:"汤者茶之司,见其沸如鱼目,微微有声,是为一沸,铫缘涌如连珠,是为二沸,腾波鼓浪,是为三沸。一沸太稚,谓之婴儿汤;三沸太老,谓之百寿汤(老汤也不可用)。若水面浮珠,声若松涛,是为第二沸,正好之候也。"《大观茶论》云:"凡用汤如鱼目、蟹眼连绎迸跃为度。"苏东坡煮茶诗:"蟹眼已过鱼眼生。"潮俗深得此法。

冲点。

取沸汤,揭罐盖,环壶口,缘壶边冲入,切忌直冲壶心,不可断续又不可迫促。铫宜提高倾注,始无涩滞之病。

Observing the Water.

In the "Tea Study[35]" it is stated: "Without hot water, how can the virtue of tea manifest?" In the "Discourse on Tea[36]" we read: "In hot water is the life of tea." When the water boils like *fish eyes*[37], sighing with a faint sound, it is the first boil; when rows of pearls surge to the edge of the kettle, it is the second boil; when it rouses like rambling waves, it is the third boil. The first boil is too young, it is called the "infant boil"; the third boil is too old, and it is called the "lifelong boil" (an old boil[38] is also not suitable for steeping);

---

[35] 茶谱 (1440), by Zhu Quan, 17th son of the Hongwu Emperor (1368–1398).

[36] 茶说 (1717), by Wang Caotang.

[37] As canonized by the afore-mentioned Lu Yu in his "Classic of Tea", it is possible to determine at which stage boiling water is optimal for brewing. If the boiling appears like surging bubbles as small as *crab-eyes*, the water temperature is lower than if they would appear as big as *fish eyes* or even pearls.

[38] Due to the chemical structure of drinking water and its benefits, some tea masters in Chaoshan prefer not to brew tea with water

when pearls float on the water surface and the sound is like the waving pines, it is the second boil, just the right steeping time. In the "Grand Treatise on Tea[39]" it is stated: "When making tea, let the water be like *fish eyes* or *crab eyes*, and let them keep gushing out in a continuous stream." In Su Dongpo's poem about brewing tea we read: "The *crab eyes* have passed, and the *fish eyes* have come.[40]" This method is an old practice in Chaozhou.

---

that has been boiled too long before brewing, or re-boiled multiple times. According to this practice, water should be poured at the right heat, but not boiled for too long, in order to preserve its original essence, or structure.

[39] 大观茶论 (1107), by the emperor Huizong (1082–1135), the only known treatise on tea written by an emperor. It is also known by the title "Tea Theory" (茶论).

[40] See note 37.

Pouring Water.

Lift the lid of the teapot and pour the boiling water in, from around the edge of the pot. Avoid pouring water directly into the center of the teapot; pour it steadily, not intermittently, and do not rush; raise the kettle properly as you pour, so no defects of astringency may be detected in the liquor.

刮沫。

冲水必使满而忌溢,满时茶沫浮白,溢出壶面,提壶盖从壶口平刮之,沫即散坠,然后盖定。

淋罐。

壶盖盖后,复以热汤遍淋壶上,以去其沫。壶外追热,则香味盈溢于壶中。

烫杯。

淋罐已毕,仍必淋杯。淋杯之汤宜直注杯心。若误触边缘,恐有破裂,俗谓烧盅热罐,方能起香。

Scraping Off the Foam.

In pouring water, it is necessary to fill the teapot without overflowing it. When the pot is full, white foam will form, floating and spilling over the pot's surface. Lift the lid and scrape it on the pot's opening to disperse the foam; then close the lid.

Drenching the Pot.

After closing the lid, pour hot water all over the pot, in order to remove the foam. This will heat up the outside of the pot, letting fragrance fill and over-infuse its body.

Heating Up the Cups.

After heating up the pot, there is no need to pour water all over the cups. To heat up each cup, pour water directly into their center. If you accidentally touch the edge, it may crack. It is said that first heating up the pot and the cups will bring forth the tea's fragrance.

洒茶。

茶叶纳后,淋罐淋杯,倾水,几番经过,正洒茶适当时候。缘洒不宜速,亦不宜迟。速则浸浸未透,香色不出,迟则香味迸出,茶色太浓,致味苦涩,全功尽废。洒则各杯轮匀,又必余沥全尽,两三洒后,覆转冲罐,俾滴尽之。 洒茶既毕,乘热,各人一杯饮之。杯缘接唇,杯面迎鼻,香味齐到,一啜而尽,三嗅杯底,味云腴,食秀美,芳香溢齿颊,甘泽润喉吻,神明凌霄汉,思想弛古今。境界至此,已得"工夫茶"三昧。

Steeping the Tea.

After the tea leaves are placed in the teapot, rinse the teapot and the cups with water, pouring water on them several times until the tea can be poured at the proper time. Pouring should be done neither too quickly nor too slowly. Too quickly, and the tea will not be steeped properly, lacking in aroma and color. Too slowly, and aroma and flavor will burst forth, making the liquor too dense, the taste bitter and astringent, wasting all your efforts. Pour the tea in a rotation, evenly between each cup, ensuring that all the tea is poured completely to the last drop. After two or three pours, rotate the teapot until it spills every drop. Once the tea is poured, each guest drinks a cup while it is still hot. Bringing the rim of the cup to the lips, the nose approaches the cup's surface, meeting its aroma; take a full sip then, and sniff the bottom of the cup three times. The flavor is rich and graceful, the aroma overflows the teeth and the cheeks, the sweetness moistens the throat and the mouth. Like deities soaring up the heavens, thoughts loosen through endless time. At this state then, one has attained the *samadhi*[41] of Gong Fu tea.

---

[41] The Sanskrit term used in Buddhism to denote meditative concentration, the attainment of "oneness", or unification between the mind and the objects of experience. In some schools of Zen Buddhism, the arts, including the tea art, are refined into the attainment of *samadhi*.

Translator's Note.

The "Chaozhou Book of Tea" belonged to an era of radical changes, when the Chinese local arts and crafts were starting to be appreciated on a wider scale, and when the area of Chaoshan had already been known for its commerce and for its people's love of tea. Weng Huidong wishes to create an essay where he can explain the basis upon which his people cultivate this art, their special customs and aesthetic preferences. The text though, is written in a style that implies its readers possess a certain knowledge of tea and traditional crafts. When he refers to specific antiques used as utensils or artefacts, he does not expound; as well as he does with quotes on tea culture by ancient authors. He merely cites those objects or works as if the reader is already familiar with their existence. These are cases where we have preferred to keep terms and names original, transcribing them according to the contemporary Pinyin system, and to add a note to the text, in order to immerse the reader in the same intellectual atmosphere the essay has been conceived in.

Other terms that have been left in the original Chinese transcription are those that are already recognized internationally in fields such as antiques appreciation or the tea art. Examples are Gong Fu tea, the Qinghua decorative style for ceramics, among others. A few additions have been made where the subjects are common, such as *bai ci* (白瓷), that is simply white porcelain in Chinese, but of the finest type that is internationally known as *blanc de chine* porcelain. In this case, the international term replaced the simple word-to-word translation, which would not convey the original characteristics of the mentioned material.

A translation of original terms has been favored instead, where the author introduces exclusive subjects and concepts that may have been intended as new to the reader, for instance while naming types of teacups, the clay textures of the pots, etc.

Overall, the style of the original text is direct, minimal yet refined; and the translation may not have been able to convey the author's elegance. Nevertheless, we hope that readers have had at least a taste of a famous art, from the time it was still encased as a local treasure; a little cup, a sip from a most precious Gong Fu teatime with a Chaoshan Master.

www.ingramcontent.com/pod-product-compliance
Lightning Source LLC
Chambersburg PA
CBHW061339040426
42444CB00011B/2992